సంఖ్యల కథ

THE NUMBER STORY

SMALL BOOK ONE

ENGLISH - TELUGU

Numbers Teach Children
Their Number Names

written and illustrated by

MISS ANNA

Early Reader Edition of *The Number Story 1*
Bronze Medal Winner, 2016 Wishing Shelf Book Award

Library of Congress Control Number: 2018902040

Names: Miss Anna, author.
Title: Number story : numbers teach children their number names / Miss Anna.
Description: Portland, OR: Lumpy Publishing, 2018.
Identifiers: ISBN 978-1-949320-19-0 | LCCN 2018902040
Summary: The pictures and rhymes present stories which introduce numbers 0-10.
Subjects: LCSH Numeration—English—Telugu--Pictorial works--Juvenile literature. | BISAC JUVENILE NONFICTION /
Languages: English—Telugu
Classification: LCC QA141.3 .M57 2018 | DDC 513—dc23

Publisher: Lumpy Publishing
Website: www.missannabooks.com
Email: missanna@missannabooks.com

Paperback: ISBN 978-1-949320-19-0
Printed in the U.S.A.   1 3 5 7 9 10 8 6 4 2

మీరు సంఖ్యల పేర్లను నేర్చుకోవాలనుకుంటున్నారా?

It is very easy and a lot of fun!

ఆది చాలా సులభంగా మరియు తమాషాగా ఉంటుంది!

Say-along our little jingle

ఈ కథను మాతో పాటు పాడండి!

starting from Number One!

మనం ఒకటవ సంఖ్య నుండి ప్రారంభిద్దము!

1

ONE looks like my one finger.

౧ ఒకటి

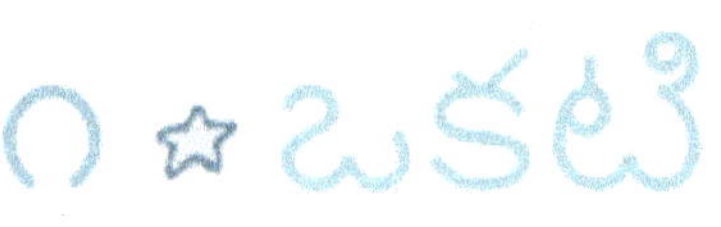

నా వెలు లాగా ఉంటుంది.

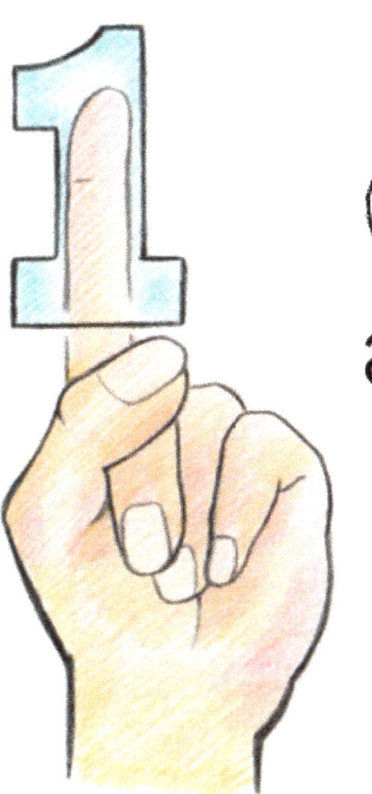

ONE!

ఒకటి!

2

TWO trails a tail.

౨ ☆ రెండు

తోకలాగా ఉంటుంది.

A TAIL! ఒక తోక!

3

THREE has bumps.

3 ☆ మూడు వంపులు ఉంటాయి.

BUMPY! వంపులు!

4

FOUR carries a sail.

ళ నలుగు

తెరచాపను మోస్తుంది.

4
A SAIL!
ఒక తెరచూప!

5

FIVE   is a racing track.

ఇ ☆ ఆయిదు

ఒక రేసింగ్ ట్రాక్.

VROOM
వువువ్!

6

SIX   curves like a snail.

౬ ఆరు

నత్తలాగా వంకరగా ఉంటుంది.

A SNAIL!
నత్త!

7

SEVEN has a sharp angle.

౭ ✫ ఏడు

ఒక పదునైన కోణం ఉంటుంది.

BE CAREFUL! IT'S SHARP!

జాగ్రత్త! ఆది పదునుగా ఉంటుంది!

EIGHT   is rollercoaster rails.

౮ ✶ ఎనిమిది

రోలర్ కోస్టర్ పట్టాలు.

YIPPEE!

NINE   is a bubble on a stick.

೯ తొమ్మిది

ఒక కర్ర మీద బుడగ.

A BUBBLE!

ఒక బుడగ!

10

TEN   is an eye of a whale.

౧౦ ⭐ పది

అది తిమింగలం కన్ను.

HELLO!
హలో!

And
మరియు
0
ZERO   is an empty pail.
౦ సున్నా
ఖాళీ బాల్చి.

IT'S EMPTY!
ఆది ఖాళీగా ఉంది!

Thank you for playing with us today.

We had a lot of fun too!

ఈరోజు మాతో ఆడినందుకు ధన్యవాదాలు.

మాకు కూడా చాలా సరదాగా ఉంది!

We are your Number friends,
Zero to Ten,
Who will be here for you~
మేము మీ సంఖ్యల మిత్రులం
సున్నా నుండి పది వరకు.
మేము ఎల్లప్పుడు మీకోసం ఇక్కడే ఉంటాము.

Bye-bye now!
See you again soon!
ఇప్పటికి బై-బై!
త్వరలో మళ్ళీ కలుద్దాము!

The Numbers are *SINGING* too!

To sing-a-long, look for Miss Anna Number Story
at your favorite music store like iTUNES.

MP3

Numbers 0-10
IDENTIFYING & COUNTING

Number Story 1 & 2

isbn: 978-0-996216-48-7

Numbers 11-20 & Ordinals
first, second, third...

Number Story 3 & 4

isbn: 978-1-945977-01-5

Numbers 0-100 & Place Values
ones, tens, hundreds...

Number Story 5 & 6

isbn: 978-1-945977-06-0

About Clocks & Telling Time
hours, minutes, seconds

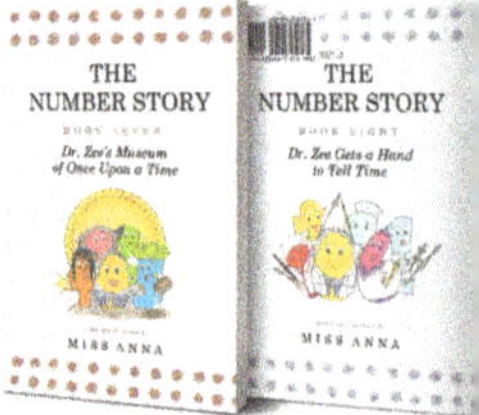

Number Story 7 & 8

isbn: 978-1-949320-40-4

For more Miss Anna books to love,
visit us at

www.missannabooks.com

Numbers are working hard all over the world!
Come Travel the World with Us!

www.ingramcontent.com/pod-product-compliance
Lightning Source LLC
Chambersburg PA
CBHW040903070726
47599CB00035B/2284